DATE DUE

FEB 1 0 2008	

DEMCO, INC. 38-2931

PLANTS OF THE RAIN FOREST

RAIN FORESTS

Lynn Stone

The Rourke Corporation, Inc.
Vero Beach, Florida 32964

Printed in the U.S.A.

PHOTO CREDITS
All photos © Lynn M. Stone

Library of Congress Cataloging-in-Publication Data

Stone, Lynn M.
    Plants of the rain forest / by Lynn M. Stone
       p. cm. — (Discovering the rain forest)
    Includes index
    ISBN 0-86593-393-6
    1. Rain forest plants—Juvenile literature. [1. Rain forest plants.
2. Rain forest ecology. 3. Ecology.]
I. Title   II. series: Stone, Lynn M.   Discovering the rain forest.
QK938.R34S88 1994
581.909'52—dc20   94-20908
                                                        CIP
                                                        AC

TABLE OF CONTENTS

PLANTS OF THE TROPICAL RAIN FOREST

The tropical rain forest is a great, green, wild garden. Tree trunks, some thick as elephants, rise from the forest floor. Shiny, oval-shaped leaves crowd branches. Vines hang and cling to bark. Palm leaves spread like fans. Moss gathers on branches and logs.

Here and there an orchid, a lily and a **bromeliad** in bloom add a dash of bright color to the green jungle. Plants are everywhere in these warm, damp forests of the **tropics**.

Rain forests are warm, wet jungles of green

NUTRIENTS

Tropical rain forest plants have special ways to survive in their wet world. Each plant, for example, has a way to find **nutrients**.

Nutrients are the healthy things that plants and animals need for growth. Think of nutrients as vitamins.

Many nutrients in the rain forest come from dead plants and animals. The process called decay releases nutrients from dead plants and animals and allows them to be re-used by living plants.

One of the important **organisms** in the decay process is the fungi family, which includes the common mushroom.

Mushrooms and other fungi help release nutrients for living plants

ROOTS AND LEAVES

The wide, shallow roots of rain forest plants are designed to take nutrients from the soil quickly. If the roots couldn't work quickly, the heavy rains would carry the nutrients away.

Leaves help plant survival, too. Most of the leaves in a tropical rain forest have sharp points called "drip tips." Water falls from drip tips easily. By drying rapidly, the leaves can avoid harmful diseases that like moisture.

The roots of tropical rain forest plants soak up water and nutrients quickly

TREES

Each of the tropical rain forests has hundreds of different kinds of trees. Their upper branches form the forest's **canopy**, or roof. A few giant trees poke their crowns, or tops, through the canopy. Some of these trees stand up to 180 feet tall.

Many trees of the rain forests have curious trunks. Some stand on a jumble of prop roots that look like spider legs. Others have odd shaped trunks, for better support.

Prop roots form a jumble of "stilts" under some rain forest trees

A tropical bromeliad blooms in a Costa Rican rain forest

A bee the color of green metal visits a jungle lily

THE CANOPY

The tropical rain forest's main canopy is a web of vines, leaves and branches about 100 feet above the ground.

The canopy is a much different place for plants and animals than the wet, dimly lit forest beneath it. Sunlight and wind easily reach the canopy, drying its branches and leaves.

Rain forest canopies are pathways in the sky for monkeys, insects, orangutans, clouded leopards, sloths and a great variety of other creatures.

The rain forest canopy is a pathway in the sky for animals

EPIPHYTES

Vines and plants called **epiphytes** cover the bark and branches of tropical rain forest trees.

Epiphytes, or "air plants," have no need for the soil of the rain forest floor. Epiphytes have special ways of living *above* the ground.

Epiphytes cling to bark with rootlike "holdfasts." The epiphytes take nutrients from plant litter that collects around their "roots."

Common epiphytes are mosses, ferns, certain orchids and bromeliads.

Long, sharp leaves of a bromeliad and other epiphytes cover a tree trunk

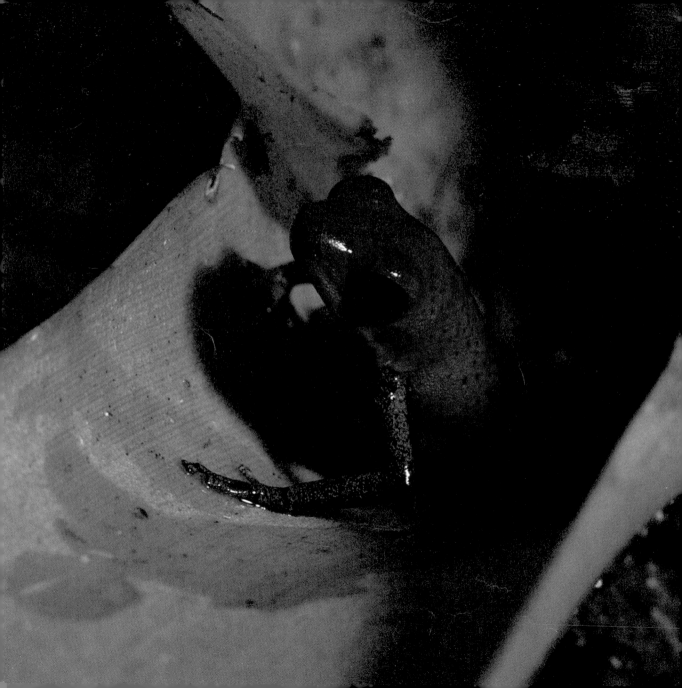

BROMELIADS

Most epiphytes take water they need from the air. Bromeliads catch water in a cuplike arrangement of leaves. During rains each bromeliad becomes a cup of water among the branches.

Every bromeliad offers a tiny **habitat**, or home, for animals high above the forest floor. Some **species**, or kinds, of frogs raise their tadpoles in bromeliad "cups." Birds drink from and bathe in bromeliad pools.

The pools are also popular with insects, spiders and other small creatures.

Poison-arrow frog peers from a bromeliad "cup"

THE TASK OF PLANTS

The plant life of the tropical rain forests supports the forests' animal life. Each animal in the rain forest either eats plants or eats other animals that eat plants.

Plants do more than feed animals. They anchor soil so that it doesn't **erode**, or wash away, into rivers.

Plants don't breathe the same way that people do. They do "process" air, however. The activity of plants helps keep rain forest temperatures steady and the air clean.

Tropical rain forest plants help keep water clean and filter air

TROPICAL RAIN FOREST PLANTS FOR PEOPLE

People have used the plants of the tropical rain forests for food and medicine for hundreds of years. Cocoa, vanilla, rubber and bamboo are products of rain forest plants.

Quinine is a medicine taken from the rain forest cinchona tree. Quinine is important in treating malaria, a serious tropical disease.

Scientists continue to find rain forest plants that will help treat human diseases.

Glossary

bromeliad (bro MILL ee ad) — a group of tropical plants that live as epiphytes and have a cuplike arrangement of leaves

canopy (KAN uh pee) — the "roof" of upper branches and leaves in a forest

epiphyte (EHP uh fite) — any of the several kinds of plants that grow on other plants, usually trees, without harming the host plant

erode (EE rode) — to eat into or wear away, especially soil or rock

habitat (HAB uh tat) — the special kind of area in which a plant or animal lives, such as the canopy of the tropical rain forest

nutrient (NU tree ent) — any of several "good" substances needed for health and growth

organism (OR gan izm) — any living thing

species (SPEE sheez) — a certain kind of plant or animal within a closely related group; for example, a *spider* lily

tropics (TRAH pihx) — a warm region of the Earth including an equal distance north and south of the equator; the region between the Tropic of Cancer and the Tropic of Capricorn on world maps

INDEX